INTERNATIONAL CENTRE FOR MECHANICAL SCIENCES

COURSES AND LECTURES - No. 24

RAYMOND D. MINDLIN
COLUMBIA UNIVERSITY, NEW YORK

POLARIZATION GRADIENT IN ELASTIC DIELECTRICS

COURSE HELD AT THE DEPARTMENT
FOR MECHANICS OF DEFORMABLE BODIES
JULY 1970

UDINE 1970

SPRINGER-VERLAG WIEN GMBH

This work is subject to copyright.

All rights are reserved,

whether the whole or part of the material is concerned

specifically those of translation, reprinting, re-use of illustrations,

broadcasting, reproduction by photocopying machine

or similar means, and storage in data banks.

© 1972 by Springer-Verlag Wien

Originally published by Springer-Verlag Wien New York in 1972

ISBN 978-3-211-81087-3   ISBN 978-3-7091-2998-2 (eBook)
DOI 10.1007/978-3-7091-2998-2

Preface

During the past decade, there has been renewed interest in efforts to extend the classical theories of continua to accommodate the effects of the structure of materials and to bridge the gap between the classical continuum theories and the dynamical theories of crystal lattices. The Born- von Karman theory of monatomic lattices of mass points has the classical theory of elasticity as its long wave limit; so that the efforts there have been to extend the connection to shorter wave lengths and to compound lattices. The modern theories of lattices of electronically polarizable atoms do not, however, have the classical continuum theory of the elastic dielectric as their long wave limit. The discrepancy lies with the continuum theory and, for monatomic lattices, can be eliminated by adding, to the stored energy of deformation and polarization, a functional dependence on the polarization gradient.

I wish to thank the Rectors and the Secretary General of the International Centre for Mechanical Sciences for their kind invitation to me to present the following series of lectures in which the derivation of the augmented continuum theory of elastic dielectrics and some of its consequences are described.

July, 1970

1. Introduction.

The linear piezoelectric effect, which occurs in elastic dielectrics, is a physical phenomenon widely used in technology. The classical continuum theory [1] incorporating the effect is about seventy-five years old and is today universally employed in the analysis and design of crystal oscillators, filters and transducers [2,3]. In systematic derivations of the classical, phenomenological theory, piezoelectricity is usually expressed as an interaction between mechanical strain and one of the electrical variables: the electric field [3] the electric displacement [4] or the electric polarization [5]. There is, however, no fundamental reason for restricting attention to the interaction of the strain, a second rank tensor, with only a first rank electrical quantity (field, displacement, polarization). It is not illogical to examine the consequences of considering an additional, linear, electromechanical effect: an interaction between the strain and, say, the polarization gradient - a second rank tensor quantity. The added complexity is justified by the fact that the resulting mathematical theory [6] has interesting novel properties: (1) it accomodates the mathematical representation of a surface energy of deformation and polarization which is absent from the classical theory but which has been measured in the laboratory [7] and calculated from atomic considerations [8];

(2) it can account [9] for an apparent anomaly observed in measurements of the electrical capacitance of thin, dielectric films [10]; (3) the additional electromechanical effect is not confined to non-centrosymmetric materials, as is the classical piezoelectric effect; (4) the resulting equations, rather than the classical ones, are the correct, long wave limit [9,11] of the equations of the modern dynamical theory of crystal lattices of electronically polarizable atoms [12,13,14,15].

2. Classical Equations.

The classical, linear theory of elastic dielectrics is expressed in terms of 25 dependent variables [3,4]:

u_i Mechanical displacement
S_{ij} Strain
T_{ij} Stress
E_i Electric field
D_i Electric displacement
P_i Polarization
φ Electric potential

These variables are linked by 25 equations:
(a) the stress-equations of motion

$$T_{ij,i} + f_j = \rho \ddot{u}_j, \qquad (2.1)$$

where ρ is the mass density and f_j is the body force;
(b) the equations of the electrostatic field

$$D_{i,i} = 0, \qquad (2.2)$$

$$E_i = -\varphi_{,i} ; \qquad (2.3)$$

(c) the strain-displacement relations

$$S_{ij} = \tfrac{1}{2}(u_{j,i} + u_{i,j}); \qquad (2.4)$$

and (d) the constitutive relations

$$T_{ij} = c^E_{ijkl} S_{kl} - e_{kij} E_k, \qquad (2.5)$$

$$D_i = e_{ikl} S_{kl} + \epsilon^S_{ik} E_k, \qquad (2.6)$$

$$P_i = D_i - \epsilon_0 E_i, \qquad (2.7)$$

where

c^E_{ijkl} = elastic stiffness (E_i = constant),
e_{ijk} = piezoelectric stress-constant
ϵ^S_{ij} = permittivity (S_{ij} = constant),
ϵ_0 = permittivity of a vacuum.

The equations of motion may be reduced to four by elimination of all the variables except u_i and φ :

$$c^E_{ijkl} u_{k,li} + e_{kij} \varphi_{,ki} + f_j = \rho \ddot{u}_j \qquad (2.8)$$

$$e_{kij} u_{i,jk} - \epsilon^S_{ij} \varphi_{,ij} = 0. \qquad (2.9)$$

3. Toupin's Variational Principle for the Classical Theory.

The classical equations of equilibrium of elastic dielectrics may be derived in an illuminating form by means of a variational principle due to Toupin [5]. A linear version of the variational principle is described in this section.

First, separate the energy density, W, of the dielectric into the stored energy density of deformation and polarization, W^L, and a remainder which is the energy density of the Maxwell electric self-field:

$$W = W^L(S_{ij}, P_i) + \frac{1}{2} \epsilon_0 \varphi_{,i} \varphi_{,i} . \tag{3.1}$$

Then define an electric enthalpy

$$H = W - E_i D_i . \tag{3.2}$$

Upon substituting (3.1), (2.7) and (2.3) in (3.2) we find

$$H = W^L(S_{ij}, P_i) - \frac{1}{2} \epsilon_0 \varphi_{,i} \varphi_{,i} + \varphi_{,i} P_i . \tag{3.3}$$

Consider a body occupying a volume V bounded by a surface S separating V from an outer vacuum V'. For such a system, the version of Toupin's principle that leads to equations equivalent to those in the preceding section, for the case of equilibrium, is

$$(3.4) \quad -\delta \int_{V^*} H \, dV + \int_V (f_i \delta u_i + E^o_i \delta P_i) dV + \int_S t_i \delta u_i \, dS = 0$$

for independent variations of u_i, P_i and φ. In (3.4), $V^* = V + V'$ and t_i is the surface traction across S and E^o_i is the external electric field. Now,

$$(3.5) \quad \delta H = \frac{\partial W^L}{\partial S_{ij}} \delta S_{ij} + \frac{\partial W^L}{\partial P_i} \delta P_i - \epsilon_0 \varphi_{,i} \delta \varphi_{,i} + \varphi_{,i} \delta P_i + P_i \delta \varphi_{,i}.$$

Define a stress T_{ij} and an effective local electric force E^L_i by

$$(3.6) \quad T_{ij} = \frac{\partial W^L}{\partial S_{ij}} \quad , \quad E^L_i = -\frac{\partial W^L}{\partial P_i}.$$

Note that, since $T_{ij} = T_{ji}$ and

$$(3.7) \quad S_{ij} = \frac{1}{2}(u_{j,i} + u_{i,j}),$$

$$T_{ij} \delta S_{ij} = T_{ij} \delta u_{j,i} = (T_{ij} \delta u_j)_{,i} - T_{ij,i} \delta u_j,$$

$$\varphi_{,i} \delta \varphi_{,i} = (\varphi_{,i} \delta \varphi)_{,i} - \varphi_{,ii} \delta \varphi,$$

$$P_i \delta \varphi_{,i} = (P_i \delta \varphi)_{,i} - P_{i,i} \delta \varphi,$$

by the chain rule of differentiation. Then

$$\delta H = -T_{ij,i} \delta u_j - (E^L_i - \varphi_{,i}) \delta P_i - (-\epsilon_0 \varphi_{,ii} + P_{i,i}) \delta \varphi$$

$$(3.8) \qquad + \left[T_{ij} \delta u_j + (-\epsilon_0 \varphi_{,i} + P_i) \delta \varphi \right]_{,i}.$$

Noting that $V^* = V + V'$ and that u_i and P_i do not exist in V', we find, after applying the divergence theorem,

$$-\delta \int_V H \, dV = \int_V \left[T_{ij,i} \delta u_j + (E_i^L - \varphi_{,i}) \delta P_i + (-\epsilon_0 \varphi_{,ii} + P_{i,i}) \delta \varphi \right] dV$$

$$- \int_S n_i \left[T_{ij} \delta u_j + (-\epsilon_0 \varphi_{,i} + P_i) \delta \varphi \right] dS , \qquad (3.9a)$$

$$-\int_{V'} H \, dV = -\int_{V'} \epsilon_0 \varphi_{,ii} \delta \varphi \, dV - \int_S \epsilon_0 n_i \varphi_{,i} \delta \varphi \, dS . \qquad (3.9b)$$

Hence, (3.4) becomes

$$\int_V \left[(T_{ij,i} + f_j) \delta u_j + (E_i^L - \varphi_{,i} + E_i^0) \delta P_i + (-\epsilon_0 \varphi_{,ii} + P_{i,i}) \delta \varphi \right] dV - \int_{V'} \epsilon_0 \varphi_{,ii} \delta \varphi \, dV$$

$$+ \int_S \left[(t_j - n_i T_{ij}) \delta u_j + n_i (\epsilon_0 [\![\varphi_{,i}]\!] - P_i) \delta \varphi \right] dS = 0 , \qquad (3.10)$$

where $[\![\varphi_{,i}]\!]$ is the jump in $\varphi_{,i}$ across S. Then the Euler equations

$$\left. \begin{array}{r} T_{ij,i} + f_j = 0 \\ E_i^L - \varphi_{,i} + E_i^0 = 0 \\ -\epsilon_0 \varphi_{,ii} + P_{i,i} = 0 \end{array} \right\} \text{ in } V, \qquad (3.11)$$

$$\varphi_{,ii} = 0 \text{ in } V' \qquad (3.12)$$

and the natural boundary conditions

$$\left. \begin{array}{r} n_i T_{ij} = t_j , \\ n_i (-\epsilon_0 [\![\varphi_{,i}]\!] + P_i) = 0, \end{array} \right\} \qquad (3.13)$$

follow from (3.10). In (3.13), $n_i(-\epsilon_0[\![\varphi_{,i}]\!]+P_i)$ is the surface charge.

The energy density of deformation and polarization is taken to be

$$(3.14) \qquad W^L = \frac{1}{2} a^s_{ij} P_i P_j + \frac{1}{2} c^P_{ijkl} S_{ij} S_{kl} + f_{kij} S_{ij} P_k$$

so that, from (3.6),

$$(3.15) \qquad \begin{cases} -E^L_j = a^s_{jk} P_k + f_{jkl} S_{kl}, \\ T_{ij} = f_{kij} P_k + c^P_{ijkl} S_{kl}. \end{cases}$$

Equations (3.7), (3.11), (3.12) and (3.15), with boundary conditions (3.13), constitute a linear version of the equations of equilibrium of elastic dielectrics in the form given by Toupin.

The relations between the new constants a^s_{ij}, f_{ijk} and c^P_{ijkl} and more familiar (or, at least, standard [16] ones) are found as follows.

From (2.3) and the second of (3.11) (omitting E^0_i as in the usual formulation) $E^L_i = -E_i$. Hence the first of (3.15) becomes

$$(3.16) \qquad E_j = a^s_{jk} P_k + f_{jkl} S_{kl}.$$

Now, the ratio of P to $\epsilon_0 E$ is the dielectric susceptibility [17]. Hence a^s_{ij} is proportional to the <u>reciprocal</u> susceptibility at constant strain (χ_{ij}^s)[*]:

$$a_{ij}^s = \epsilon_0^{-1} \chi_{ij}^s. \qquad (3.17)$$

With (3.17), (3.16) becomes

$$E_j = \epsilon_0^{-1} \chi_{jk}^s P_k + f_{jkl} S_{kl}. \qquad (3.18)$$

Define susceptibility at constant strain, η_{ij}^s, according to

$$\eta_{ij}^s \chi_{jk}^s = \delta_{ik} \qquad (3.19)$$

and multiply both sides of (3.18) by $\epsilon_0 \eta_{mj}$ to get

$$\epsilon_0 \eta_{mj}^s E_j = P_m + \epsilon_0 \eta_{mj}^s f_{jkl} S_{kl}. \qquad (3.20)$$

Then eliminate P between (3.20) and (2.7), with the result

$$D_i = - \epsilon_0 \eta_{ij}^s f_{jkl} S_{kl} + \epsilon_0 (\delta_{ij} + \eta_{ij}^s) E_j. \qquad (3.21)$$

Accordingly, comparing (3.21) with (2.6),

$$e_{ikl} = - \epsilon_0 \eta_{ij}^s f_{jkl}, \quad \text{or} \quad f_{jkl} = - \epsilon_0^{-1} \chi_{ij}^s e_{ikl}; \qquad (3.22)$$

$$e_{ij}^s = \epsilon_0 (\delta_{ij} + \eta_{ij}^s), \quad \text{or} \quad \eta_{ij}^s = \epsilon_0^{-1} \epsilon_{ij}^s - \delta_{ij}. \qquad (3.23)$$

To find c_{ijkl}^P in terms of c_{ijkl}^E, first substitute the expression for P_i, given in terms of E_i and S_{ij} in (3.20),

---

[*] Note that, according to the IRE Standards on Piezoelectric Crystals [16] reciprocal susceptibility is designated by χ whereas many text-books, e.g. [17] employ χ for susceptibility, In the IRE Standards, the susceptibility is η, following Voigt [1].

into the second of (3.15):

$$T_{ij} = f_{mij}(\epsilon_0 \eta^s_{mk} E_k - \epsilon_0 \eta^s_{mn} f_{nkl} S_{kl}) + c^P_{ijkl} S_{kl}$$

or, from (3.22),

(3.24) $\quad T_{ij} = (c^P_{ijkl} - \epsilon_0 \eta^s_{mn} f_{mij} f_{nkl}) S_{kl} - e_{kij} E_k$.

Comparing (3.24) with (2.5), we see that

(3.25) $\quad c^E_{ijkl} = c^P_{ijkl} - \epsilon_0 \eta^s_{mn} f_{mij} f_{nkl}$

or, from (3.22),

(3.26) $\quad c^P_{ijkl} = c^E_{ijkl} + \epsilon_0^{-1} \chi^s_{mn} e_{mij} e_{nkl}$.

With the aid of the foregoing expressions for the constants a^s_{ij}, c^P_{ijkl} and f_{ijk}, we may readily reduce the three equations of equilibrium (3.11) and the two constitutive equations (3.15) to the classical form of equations on u_i and φ given at the end of the preceding section - by elimination of P_i. First, to solve the first of (3.15) for P_i, recall that the reciprocal of a^s_{ij} is $\epsilon_0 \eta^s_{ij}$. Then, upon multiplication, we have

(3.27) $\quad P_i = -\epsilon_0 \eta^s_{ij}(E^L_j + f_{jkl} S_{kl})$.

Substitute this for P_i in the second of (3.15) to obtain

$$T_{ij} = -f_{kij} \epsilon_0 \eta^s_{km}(E^L_m + f_{mpq} S_{pq}) + c^P_{ijkl} S_{kl}.$$

From the second equilibrium equation $E^L_i = \varphi_{,i}$, again omitting E^0_i.

Reduction to Classical Field Equations

Also $S_{ij} = \frac{1}{2}(u_{j,i} + u_{i,j})$. Hence

$$T_{ij,i} = \left(c^P_{ijkl} - \epsilon_0 \eta^S_{mn} f_{mij} f_{nkl}\right) u_{k,li} - \epsilon_0 \eta^S_{km} f_{kij} \varphi_{,mi}.$$

Then, from the first of (3.22) and from (3.25),

$$T_{ij,i} + f_j = c^E_{ijkl} u_{k,li} + e_{kij} \varphi_{,ki} + f_j$$

which is the left hand side of (2.8), as required.

Next, from (3.27),

$$P_{i,i} = -\epsilon_0 \eta^S_{ij} \left(E^L_{j,i} + f_{jkl} S_{kl,i}\right)$$

$$= -\epsilon_0 \eta^S_{ij} \varphi_{,ij} - \epsilon_0 \eta^S_{ij} f_{jkl} u_{k,li}.$$

Hence, the left hand side of the last of the equilibrium equations (3.11) becomes

$$-\epsilon_0 \varphi_{,ii} + P_{i,i} = -\epsilon_0 \eta^S_{ij} f_{jkl} u_{k,li} - \epsilon_0 (\delta_{ij} + \eta^S_{ij}) \varphi_{,ij}$$

or, from the first of (3.22) and (3.23),

$$-\epsilon_0 \varphi_{,ii} + P_{i,i} = e_{ikl} u_{k,li} - \epsilon^S_{ij} \varphi_{,ij}$$

which is the left hand side of (2.9), as required.

4. Polarization Gradient.

Toupin's form of the classical equations of elastic dielectrics reveals an omission in the classical theory. His equation

$$E_i^L - \varphi_{,i} + E_i^0 = 0$$

is the "equation of intramolecular force balance" [5] which he derived from fundamental considerations of the equilibrium of electrical forces, but which does not appear in the usual formulations. Granted the validity of the equation, it is significant that no boundary condition is associated with it. Whereas there is an equilibrium equation associated with each of the variables u_i, φ and P_i, only the variables u_i and φ are accompanied by boundary conditions. There is no coefficient of δP_i in the surface integral in (3.10) to complement that in the volume integral. This lack can be traced back to the absence of a functional dependence of W^L on the polarization gradient $P_{j,i}$. In fact, if we were to start by assuming dependence of W^L on the displacement and polarization and their gradients and truncate after the first gradient, $P_{j,i}$ would remain. Only u_i and the antisymmetric part of $u_{j,i}$ would have to be discarded - on the grounds of required translational and rotational invariance of W^L. The only justification for discarding $P_{j,i}$ would be its possible lack of importance judged on practical considerations. However, as we shall see in Sections 8 and 9, inclusion of $P_{j,i}$ in W^L results in the extension of the classical theory to accomodate interesting phenomena which are observed in nature but are not included in the classical theory. Furthermore, as shown in Section 10, the polarization gradient represents the continuum approximation to certain interatomic interactions which are known to contribute to fitting the predictions of crystal lattice theories to the results of neutron diffraction experiments.

To extend Toupin's variational principle to account for the contribution of the polarization gradient, it is only necessary to replace (3.3) with

$$H = W^L(S_{ij}, P_i, P_{j,i}) - \frac{1}{2}\epsilon_0 \varphi_{,i} \varphi_{,i} + \varphi_{,i} P_i. \quad (4.1)$$

In addition, we shall include the kinetic energy so that (3.4) becomes

$$\delta \int_{t_0}^{t_1} dt \int_{V^*} (\frac{1}{2}\rho \dot{u}_i \dot{u}_i - H) dV + \int_{t_0}^{t_1} dt \left[\int_V (f_i \delta u_i + E_i^0 \delta P_i) dV + \int_S t_i \delta u_i dS \right] = 0, \quad (4.2)$$

for independent variations of u_i, φ and P_i between fixed limits at times t_0 and t_1.

The only additional operations involved are a new definition

$$E_{ij} = \frac{\partial W^L}{\partial P_{j,i}} \quad (4.3)$$

and two integrations:

$$\tfrac{1}{2}\delta\int_{t_0}^{t_1}\dot u_i\dot u_i\,dt = \int_{t_0}^{t_1}\dot u_i\delta\dot u_i\,dt = \dot u_i\delta u_i\Big]_{t_0}^{t_1} - \int_{t_0}^{t_1}\ddot u_i\delta u_i\,dt = -\int_{t_0}^{t_1}\ddot u_i\delta u_i\,dt,$$

$$\int_V E_{ij}\delta P_{j,i}\,dV = \int_V\left[(E_{ij}\delta P_j)_{,i} - E_{ij,i}\delta P_j\right]dV = \int_S n_i E_{ij}\delta P_j\,dS - \int_V E_{ij,i}\delta P_j\,dV.$$

With these results and (3.9), (4.2) becomes

$$\int_{t_0}^{t_1}dt\int_V\left[(T_{ij,i}+f_j-\rho\ddot u_j)\delta u_j + (E_{ij,i}+E_j^L-\varphi_{,j}+E_j^0)\delta P_j\right.$$
$$\left.+(-\epsilon_0\varphi_{,ii}+P_{i,i})\delta\varphi\right]dV - \int_{t_0}^{t_1}dt\int_{V'}\epsilon_0\varphi_{,ii}\,\delta\varphi\,dV$$

(4.4) $$+\int_{t_0}^{t_1}dt\int_S\left[(t_j-n_i T_{ij})\delta u_j - n_i E_{ij}\delta P_j + n_i(\epsilon_0[\![\varphi_{,i}]\!]-P_i)\delta\varphi\right]dS=0.$$

Then the Euler equations are

(4.5) $\text{in } V$ $$\begin{cases} T_{ij,i}+f_j = \rho\ddot u_j, \\ E_{ij,i}+E_j^L-\varphi_{,j}+E_j^0 = 0, \\ -\epsilon_0\varphi_{,ii}+P_{i,i} = 0; \end{cases}$$

(4.6) $$\varphi_{,ii} = 0 \quad \text{in } V',$$

with natural boundary conditions

(4.7) $$\begin{cases} n_i T_{ij} = t_j, \\ n_i E_{ij} = 0, \\ n_i(-\epsilon_0[\![\varphi_{,i}]\!]+P_i) = 0. \end{cases}$$

Constitutive Equations

For W^L, we take

$$W^L = b^o_{ij} P_{j,i} + \frac{1}{2} a^{SG}_{ij} P_i P_j + \frac{1}{2} b^{PS}_{ijkl} P_{j,i} P_{l,k} + \frac{1}{2} c^{PG}_{ijkl} S_{ij} S_{kl}$$
$$+ d^P_{ijkl} P_{j,i} S_{kl} + f^G_{ijk} P_i S_{jk} + i^S_{ijk} P_i P_{k,j}. \qquad (4.8)$$

The superscripts S, P and G, designating strain, polarization and polarization gradient, respectively, will be omitted in the sequel where no ambiguity results.

Substituting (4.8) in (3.6) and (4.3), we find

$$\left.\begin{aligned}
-E^L_j &= a_{jk} P_k + d_{jkl} P_{l,k} + f_{jkl} S_{kl}, \\
E_{ij} &= d_{kij} P_k + b_{ijkl} P_{l,k} + d_{ijkl} S_{kl} + b^o_{ij}, \\
T_{ij} &= f_{kij} P_k + d_{klij} P_{l,k} + c_{ijkl} S_{kl}.
\end{aligned}\right\} \qquad (4.9)$$

The field equations (4.5) and (4.6), the constitutive equations (4.9) and the strain displacement relation $S_{ij} = \frac{1}{2}(u_{j,i} + u_{i,j})$ comprise the equations of the augmented theory [6]. It is apparent, from the form of the integrand of the surface integral in (4.4), what boundary conditions other than (4.7) are admissible. Thus, in place of the traction $n_i T_{ij}$, may be substituted the displacement u_i or a component of either and the resultant of the other in the plane at right angles. The same possibilities are open for $n_i E_{ij}$ and the polarization P_i. Finally either the surface charge $n_i(-\epsilon_o [\![\varphi_{,i}]\!] + P_i)$ or the potential φ

may be specified. Of prime importance is the fact that both the potential φ and the polarization P_i may be specified. This is a latitude not permissible in the classical theory.

5. Centrosymmetric Materials: Cubic.

One of the novel properties of the augmented theory of elastic dielectrics is its accomodation of an electromechanical interaction even for materials with centrosymmetric physical properties. This may be seen by inspection of the energy density W^L as given in (4.8). For centrosymmetry, $f_{ijk} = \dot{d}_{ijk} = 0$, since there are no centrosymmetric tensors of odd rank. However, this still leaves d_{ijkl} which is the coefficient of an electromechanical coupling term in the augmented theory but does not appear in the classical theory.

As a simple example of crystallographic centrosymmetry, consider the cubic point group $m3m$ (or O_h) the generators for which are [19]

$$\begin{pmatrix} -1 & 0 & 0 \\ 0 & -1 & 0 \\ 0 & 0 & -1 \end{pmatrix} \begin{pmatrix} 0 & 0 & 1 \\ 1 & 0 & 0 \\ 0 & 1 & 0 \end{pmatrix} \begin{pmatrix} 1 & 0 & 0 \\ 0 & 0 & 1 \\ 0 & -1 & 0 \end{pmatrix} \quad (5.1)$$

Applying these to the coefficients in (4.8), we find

$$f_{ijk} = 0, \quad \dot{d}_{ijk} = 0,$$
$$a_{ij} = a_{11}\delta_{ij}, \quad b^o_{ij} = b_o \delta_{ij}, \quad (5.2a)$$

$$b_{ijkl} = b\delta_{ijkl} + b_{12}\delta_{ij}\delta_{kl} + b_{44}(\delta_{ik}\delta_{jl}+\delta_{il}\delta_{jk}) + b_{77}(\delta_{ik}\delta_{jl}-\delta_{il}\delta_{jk}),$$

(5.2b)
$$c_{ijkl} = c\delta_{ijkl} + c_{12}\delta_{ij}\delta_{kl} + c_{44}(\delta_{ik}\delta_{jl}+\delta_{il}\delta_{jk}),$$

$$d_{ijkl} = d\delta_{ijkl} + d_{12}\delta_{ij}\delta_{kl} + d_{44}(\delta_{ik}\delta_{jl}+\delta_{il}\delta_{jk}),$$

where the coefficients $a_{11}, b_0, b_{pq}, c_{pq}$ and d_{pq} are constants and

(5.3)
$$b = b_{11} - b_{12} - 2b_{44},$$
$$c = c_{11} - c_{12} - 2c_{44},$$
$$d = d_{11} - d_{12} - 2d_{44}.$$

In (5.2), δ_{ij} is the Kronecker delta and δ_{ijkl} is unity if all its indices are alike but zero otherwise.

Insertion of (5.2) in the constitutive equations (4.9) reduces the latter to

$$-E_i^L = a_{11} P_i,$$

(5.4)
$$E_{ij} = b\delta_{ijkl} P_{l,k} + b_{12}\delta_{ij} P_{k,k} + b_{44}(P_{j,i}+P_{i,j}) + b_{77}(P_{j,i}-P_{i,j})$$
$$+ d\delta_{ijkl} S_{kl} + d_{12}\delta_{ij} S_{kk} + 2d_{44} S_{ij} + b_0 \delta_{ij},$$

$$T_{ij} = d\delta_{ijkl} P_{l,k} + d_{12}\delta_{ij} P_{k,k} + d_{44}(P_{j,i}+P_{i,j})$$
$$+ c\delta_{ijkl} S_{kl} + c_{12}\delta_{ij} S_{kk} + 2c_{44} S_{ij}.$$

Equations of Motion: Cubic Symmetry

The "displacement" equations of motion are obtained by substituting (5.4) into (4.5) and employing $S_{ij} = \frac{1}{2}(u_{j,i} + u_{i,j})$ with the result

$$c\delta_{ijkl}u_{l,ki} + c_{12}u_{k,kj} + c_{44}(u_{j,ii} + u_{i,ji}) +$$

$$+ d\delta_{ijkl}P_{l,ki} + d_{12}P_{k,kj} + d_{44}(P_{j,ii} + P_{i,ji}) + f_j = \rho\ddot{u}_j,$$

$$d\delta_{ijkl}u_{l,ki} + d_{12}u_{k,kj} + d_{44}(u_{j,ii} + u_{i,ji}) - a_{11}P_j - \varphi_{,j} + \quad (5.5)$$

$$+ b\delta_{ijkl}P_{l,ki} + b_{12}P_{k,kj} + b_{44}(P_{j,ii} + P_{i,ji}) + b_{77}(P_{j,ii} - P_{i,ji}) + E_j^0 = 0,$$

$$-\epsilon_0\varphi_{,ii} + P_{i,i} = 0.$$

6. Centrosymmetric Materials: Isotropic.

For centrosymmetric isotropic materials it is only necessary to set

(6.1) $\quad b_{11} = b_{12} + 2b_{44}, \quad c_{11} = c_{12} + 2c_{44}, \quad d_{11} = d_{12} + 2d_{44};$

so that

(6.2) $\quad\quad\quad\quad b = c = d = 0$

in the equations for the centrosymmetric cubic case. The equations of motion (5.5) then become, in vector notation,

(6.3)
$$c_{44}\nabla^2\underline{u} + (c_{12}+c_{44})\underline{\nabla}\,\underline{\nabla}\cdot\underline{u} + d_{44}\nabla^2\underline{P} + (d_{12}+d_{44})\underline{\nabla}\,\underline{\nabla}\cdot\underline{P} + \underline{f} = \rho\underline{\ddot{u}},$$
$$d_{44}\nabla^2\underline{u} + (d_{12}+d_{44})\underline{\nabla}\,\underline{\nabla}\cdot\underline{u} + (b_{44}+b_{77})\nabla^2\underline{P} + (b_{12}+b_{44}-b_{77})\underline{\nabla}\,\underline{\nabla}\cdot\underline{P} - a_{11}\underline{P} - \underline{\nabla}\varphi + \underline{E}^0 = 0,$$
$$-\epsilon_0\nabla^2\varphi + \underline{\nabla}\cdot\underline{P} = 0.$$

Schwartz [18] has given a complete solution of (6.3) in terms of two vector functions, a scalar function and the potential φ:

(6.4)
$$\underline{u} = \underline{B} - \tfrac{1}{2}(1-k)\underline{\nabla}(\underline{r}\cdot\underline{B} + B_0) + a_{11}^{-1}c_{44}k_2(k_2-k_1)\underline{\nabla}\,\underline{\nabla}\cdot\underline{B} - k_2(\underline{K} - \ell_2^2\underline{\nabla}\,\underline{\nabla}\cdot\underline{K})$$
$$-\epsilon_0 k_1\underline{\nabla}\varphi + a_{11}^{-1}k_2(1+a_{11}\epsilon_0)(1-\ell_1^2\nabla^2)\underline{\nabla}\varphi,$$

$$\underline{P} = -a_{11}^{-1}c_{44}(k_2-k_1)\underline{\nabla}\,\underline{\nabla}\cdot\underline{B} + \epsilon_0\underline{\nabla}\varphi - a_{11}^{-1}(1+a_{11}\epsilon_0)(1-\ell_1^2\nabla^2)\underline{\nabla}\varphi + \underline{K} - \ell_2^2\underline{\nabla}\,\underline{\nabla}\cdot\underline{K},$$

where $\underset{\sim}{B}, \underset{\sim}{B_0}, \underset{\sim}{K}$ and φ must satisfy the equations

$$c_{44} \nabla^2 \underset{\sim}{B} = -\underset{\sim}{f},$$

$$c_{44} \nabla^2 B_0 = \underset{\sim}{r} \cdot \underset{\sim}{f},$$

$$a_{11}(1 - \ell_2^2 \nabla^2) \underset{\sim}{K} = \underset{\sim}{E}^0 - k_2 \underset{\sim}{f}, \qquad (6.5)$$

$$(1 + a_{11}\epsilon_0)(1 - \ell_1^2 \nabla^2)\nabla^2 \varphi = \nabla \cdot \underset{\sim}{E}^0 - k_1 \underset{\sim}{\nabla} \cdot \underset{\sim}{f},$$

in which $\underset{\sim}{r}$ is the position vector and

$$k = c_{44}/c_{11}, \quad k_1 = d_{11}/c_{11}, \quad k_2 = d_{44}/c_{44}, \qquad (6.6)$$

$$\ell_1^2 = (b_{11}c_{11} - d_{11}^2)/c_{11}(a_{11} + \epsilon_0^{-1}), \quad \ell_2^2 = \left[(b_{44} + b_{77})c_{44} - d_{44}^2\right]/a_{11}c_{44}. \qquad (6.7)$$

Among other solutions, Schwartz gives the fundamental one for the concentrated force $\underset{\sim}{F}$:

$$\underset{\sim}{B} = \frac{\underset{\sim}{F}}{4\pi c_{44} r}, \qquad B_0 = 0, \qquad (6.8)$$

$$\underset{\sim}{K} = -\frac{k_2 e^{-r/\ell_2}}{4\pi a_{11} \ell_2^2 r} \underset{\sim}{F}, \quad \varphi = \frac{k_1}{4\pi(1+\epsilon_0 a_{11})} \underset{\sim}{F} \cdot \underset{\sim}{\nabla}\left(\frac{1 - e^{-r/\ell_1}}{r}\right). \qquad (6.9)$$

7. Surface Energy of Deformation and Polarization.

It will be observed that the internal energy density of polarization and deformation, W^L, contains the linear term, $b^0_{ij} P_{j,i}$, in the polarization gradient. The analogous term linear in the strain, say $c^0_{ij} S_{ij}$, is omitted as it is of no consequence. It leads to a homogeneous stress which, in a bounded body with a free surface, can be removed by a homogeneous strain which, in turn, can be regarded as the reference configuration. The situation is otherwise for the term linear in the polarization gradient. Although this, too, produces a homogeneous field, it is a field of E_{ij} rather than T_{ij}. The removal of $n_i E_{ij}$ to free a surface results, as is illustrated in Section 8, in a polarization and strain which decay exponentially into the interior of the body.

To separate a body into two parts along a surface, a <u>bond</u> energy must be overcome. This is the energy that would be required to break the atomic bonds across the surface if the strain and polarization were <u>prevented</u> from developing, say by some hypothetical external field. The <u>release</u> of such a field would result in a deformation and polarization, localized near the surface, with which is associated a surface energy of deformation and polarization – always negative. Thus, the energy required to separate a body into two parts along a surface is the bond energy less the absolute value of the surface energy of deformation and polarization. The former is not represented in the present theory, but the latter is introduced via the linear term $b^0_{ij} P_{j,i}$ in W^L and a formula for that energy may be found as follows.

The total energy in the system in equilibrium is

$$\int_{V^*} W\, dV = \int_{V^*} \left(W^L + \tfrac{1}{2}\epsilon_0 \varphi_{,i}\varphi_{,i}\right) dV = \int_V W^L dV + \int_{V^*} \tfrac{1}{2}\epsilon_0 \varphi_{,i}\varphi_{,i}\, dV. \qquad (7.1)$$

By means of the constitutive equations (4.9), W^L may be expressed in the form

$$W^L = \tfrac{1}{2} T_{ij} S_{ij} + \tfrac{1}{2} E_{ij} P_{j,i} - \tfrac{1}{2} E^L_i P_i + \tfrac{1}{2} b^0_{ij} P_{j,i}. \qquad (7.2)$$

Through the use of the chain rule of differentiation and the divergence theorem,

$$\int_V T_{ij} S_{ij}\, dV = \int_V T_{ij} u_{j,i}\, dV = \int_V \left[(T_{ij} u_j)_{,i} - T_{ij,i} u_j\right] dV =$$

$$= \int_S n_i T_{ij} u_j\, dS - \int_V T_{ij,i} u_j\, dV,$$

$$\int_V E_{ij} P_{j,i}\, dV = \int_V \left[(E_{ij} P_j)_{,i} - E_{ij,i} P_j\right] dV = \int_S n_i E_{ij} P_j\, dS - \int_V E_{ij,i} P_j\, dV,$$

$$\int_V b_{ij} P_{j,i}\, dV = \int_S n_i b^0_{ij} P_j\, dS,$$

$$\int_{V^*} \varphi_{,i}\varphi_{,i}\, dV = \int_{V^*} \left[(\varphi_{,i}\varphi)_{,i} - \varphi_{,ii}\varphi\right] dV =$$

$$= \int_S n_i [\![\varphi_{,i}]\!]\varphi\, dS - \int_V \varphi_{,ii}\varphi\, dV - \int_{V'} \varphi_{,ii}\varphi\, dV.$$

Upon assembling these results into (7.1), we have

$$(7.3) \quad \int_{V^*} W\,dV = \tfrac{1}{2}\int_S n_i\left[b^o_{ij}P_j + T_{ij}u_j + E_{ij}P_j + \epsilon_o[\![\varphi_{,i}]\!]\varphi\right]dS - \tfrac{1}{2}\int_V (T_{ij,i}u_j + E_{ij,i}P_j + E^L_j P_j + \epsilon_o \varphi_{,ii}\varphi)dV - \tfrac{1}{2}\int_{V'}\epsilon_o\varphi_{,ii}\varphi\,dV.$$

Application of the equilibrium equations, (4.5) and (4.6) with $\rho\ddot{u}_j = 0$, to (7.3) reduces it to

$$(7.4) \quad \int_{V^*} W\,dV = \tfrac{1}{2}\int_V (f_j u_j + E^o_j P_j)dV + \tfrac{1}{2}\int_S n_i\left[T_{ij}u_j + E_{ij}P_j - (-\epsilon_o[\![\varphi_{,i}]\!]+P_i)+b^o_{ij}P_j\right]dS.$$

Then, in the absence of external fields $(f_j = E_j = 0)$ and with the boundary free, i.e. (4.7) with $t = 0$, (7.4) reduces to

$$(7.5) \quad \int_{V^*} W\,dV = \tfrac{1}{2}\int_S n_i b^o_{ij} P_j\,dS.$$

This is the energy of deformation and polarization which must be added to the bond energy to obtain the total energy required to separate the material into two parts along the surface S. As indicated by Schwartz [18], the additional energy is always negative as a consequence of the positive-definiteness of the quadratic part of the total energy which, in turn, is required for stability – i.e. because energy must be stored, rather than generated, during the application of external body and surface forces. Thus:

$$(7.6) \quad \begin{aligned}\int_{V^*} W\,dV &= \int_{V^*}\left[\tfrac{1}{2}\varphi_{,i}\varphi_{,i} + (W^L - b^o_{ij}P_{j,i})\right]dV + \int_V b^o_{ij}P_{j,i}\,dV, \\ &= \int_{V^*}\left[\tfrac{1}{2}\varphi_{,i}\varphi_{,i} + (W^L - b^o_{ij}P_{j,i})\right]dV + \int_S n_i b^o_{ij}P_j\,dS.\end{aligned}$$

Combining (7.5) and (7.6):

$$\int_{V^*} \left[\frac{1}{2}\varphi_{,i}\varphi_{,i} + (W^L - b^0_{ij}P_{j,i})\right]dV + \frac{1}{2}\int_S n_i b_{ij} P_j \, dS = 0. \quad (7.7)$$

But the volume integral in (7.7) is greater than zero. Hence

$$\int_S n_i b^0_{ij} P_j \, dS < 0. \quad (7.8)$$

Finally, we may define

$$T = \frac{1}{2}\left[n_i b^0_{ij} P_j\right]_S \quad (7.9)$$

as the surface energy of deformation and polarization per unit area, sometimes called the surface tension.

8. Energy at a Plane, Free Surface.

In this section the solution is given for the displacement, polarization, potential and surface energy of deformation and polarization in the half-space of a centrosymmetric cubic crystal bounded by a free (100) face. The same solution applies to the isotropic half-space if the conditions (6.1) are applied.

For this problem, the fields are one-dimensional and in equilibrium so that, for the half-space $x \geq 0$, the equations of motion (5.5) reduce to

(8.1)
$$c_{11} \partial^2 u + d_{11} \partial^2 P = 0,$$
$$d_{11} \partial^2 u + b_{11} \partial^2 P - a_{11} P_1 - \partial \varphi = 0,$$
$$-\epsilon_0 \partial^2 \varphi + \partial P = 0;$$

and the boundary conditions (4.7), on $x = 0$, become

(8.2)
$$c_{11} \partial u + d_{11} \partial P = 0,$$
$$d_{11} \partial u + b_{11} \partial P = -b_0,$$
$$-\epsilon_0 \partial \varphi + P = 0,$$

where $\partial = d/dx$ and u, P are the x-components of u_i and P_i.

Solution of Equations for a Free Half-Space

Consider

$$u = A_1 e^{-x/\ell},$$

$$P = A_2 e^{-x/\ell}, \qquad (8.3)$$

$$\varphi = A_3 e^{-x/\ell}.$$

Upon substituting (8.3) into (8.1) we find

$$A_3 = -\ell A_2/\epsilon_0 = \ell c_{11} A_1/\epsilon_0 d_{11}, \qquad (8.4)$$

$$\ell = \left(\frac{b_{11} c_{11} - d_{11}^2}{c_{11}(a_{11}+\epsilon_0^{-1})}\right)^{\frac{1}{2}}; \qquad (8.5)$$

i.e., ℓ is the same as ℓ_1 in (6.7). Positive-definiteness of W^L requires ℓ to be real.

With (8.4), the first and third of the boundary conditions (8.2) are satisfied identically and the second boundary condition yields

$$A_1 = -\frac{b_0 d_{11}}{c_{11} \ell (a_{11}+\epsilon_0^{-1})}. \qquad (8.6)$$

Then, from (8.4),

$$A_2 = \frac{b_0}{\ell (a_{11}+\epsilon_0^{-1})}, \quad A_3 = -\frac{b_0}{1+a_{11}\epsilon_0^{-1}}. \qquad (8.7)$$

It may be seen that the freeing of the boundary results in a field of displacement, polarization and potential localized at the surface, decaying into the interior with decay constant ℓ. Associated with the localized field is a surface energy of deformation and polarization, per unit area, given by (7.9):

$$(8.8) \qquad T = - \frac{b_0^2}{2 \ell a_{11}(1+a_{11}^{-1}\epsilon_0^{-1})}.$$

Note that, from (3.17), (3.19) and (3.23), $1+a_{11}^{-1}\epsilon_0^{-1} = \epsilon_{11}^s/\epsilon_0$, i.e. the dielectric constant in the [100] direction.

Askar, Lee and Cakmak [11] have calculated a_{11}, b_{11}, c_{11} and d_{11} for several alkali halide crystals by relating them to analogous constants in a Cochran [13] lattice of Dick--Overhauser [12] shell-model atoms in an NaCl structure. The correspondence is not complete because the NaCl lattice is diatomic whereas the continuum is a limiting form of a monatomic lattice. However, the calculation at least yields order of magnitude results. For NaCl, for example, they find

$$(8.9) \qquad \ell = 1.3 \times 10^{-8} cm, \quad T = -59 \text{ erg}/cm^2.$$

The decay constant ℓ is about half the distance between nearest neighbor atoms in the NaCl crystal, indicating extremely rapid decay away from the surface. The surface energy of deformation and polarization is about thirty percent of the bond energy, so that it is far from negligible. Both of these results conform to experimental observations [24, 7].

Additional solutions for surface energies of deformation and polarization in isotropic materials, have been obtained by Schwartz [18], for internal and external spherical surfaces, and by Askar, Lee and Cakmak [25] for internal spherical and cylindrical surfaces and a crack. The latter have used average values of their previous calculations of constants for alkali halides [11] to obtain some quantitative results. They find that the absolute values of T are reduced by curvature of the surface and they also conclude that T is finite at the root of a crack.

9. Capacitance of Thin, Dieletric Films.

As a second example, consider the capacitance of a metal-dielectric-metal sandwich. Suppose the middle plane of the structure is the plane of y and z. If end effects are neglected, the fields are one-dimensional and the equations of equilibrium are given by (8.1). Also, assume that the conditions of isotropy (6.1) hold.

Let us first solve the problem within the framework of the classical theory. Then b_{11} and d_{11} are zero and the equations of equilibrium reduce to

(9.1)
$$c_{11} \partial^2 u = 0,$$
$$a_{11} P + \partial \varphi = 0,$$
$$-\epsilon_0 \partial^2 \varphi + \partial P = 0.$$

For a dielectric layer with traction-free surfaces at $x \pm h$, on which are impressed voltages $\pm V$, the boundary conditions, according to the classical theory, are

(9.2)
$$(\partial u)_{x = \pm h} = 0,$$
$$(\varphi)_{x = \pm h} = \pm V;$$

and the solution of (9.1), subject to these boundary conditions, is, except for additive constants in u and φ,

$$u = 0,$$
$$P = -\epsilon_0 \eta V h, \qquad (9.3)$$
$$\varphi = V x / h,$$

where $\eta = (\epsilon_0^{-1} a_{11}^{-1})$ is the dielectric susceptibility.

The capacitance (per unit area) is the ratio of the surface charge (per unit area) to the voltage drop across the layer:

$$C = \frac{(\epsilon_0 \partial \varphi - P)_{x=\pm h}}{2V} = \frac{\epsilon_0(1+\eta)}{2h} = \frac{\epsilon}{2h}, \qquad (9.4)$$

where ϵ is the permittivity of the dielectric.

Thus, according to the classical theory, there is no strain, the polarization is uniform, the potential varies linearly through the thickness and the capacitance is inversely proportional to the thickness, so that a graph of inverse capacitance vs. thickness is a straight line through the origin. However, in a series of experiments with a variety of very thin dielectric films between metal electrodes, Mead [20,21,22] found a different relation between inverse capacitance and thickness. His experimental data fall on straight lines which, if extended to zero thickness, have positive intercepts of inverse capacitance, as illustrated in Fig.9.1. Initially [20], Mead suggested that the anomaly might be due to penetration of the electric field into the electrodes; but subsequently [21,22] he abandoned that view, although it has been supported by Ku and Ullman [23].

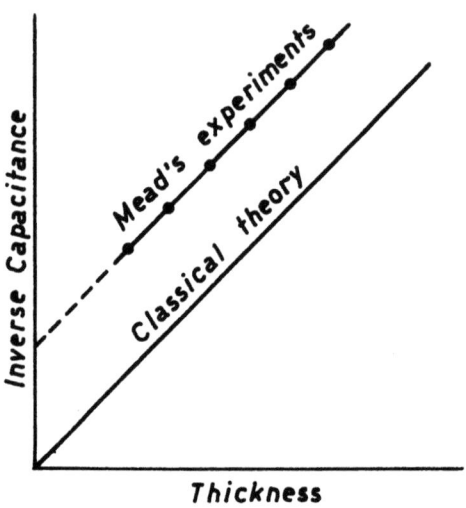

Fig.9.1

An alternative explanation is found in the augmented theory of elastic dielectrics.

The solution of the augmented equations of equilibrium (8.1), analogous to the one just given for the classical equations, requires a boundary condition in addition to the surface traction and potential. The additional condition can be the specification of the polarization at the surface of the dielectric. Now, the polarization at a boundary of the dielectric in a metal-dielectric-metal sandwich will depend on the physical properties of the adjacent electrode and metal-dielectric interface; and these properties are outside the compass of the theory of dielectrics. However, since the polarization in the metal is zero, it is reasonable to assume that the surface polarization in the dielectric, if not actually zero, will lie between zero and the classical value given by the second of (9.3). Thus, assuming that the two electrodes and interfaces are the same, their influence on the surface polarization may be introduced, phenomenologically, by setting the boundary condition

$$(P)_{x=\pm h} = -k\epsilon_0 \eta V/h, \qquad 0 \le k \le 1. \tag{9.5a}$$

The classical condition is $k=1$ while $k=0$ describes continuity of polarization across the interfaces. We suppose also, as is assumed in the classical theory, that the traction across the interfaces is zero. From (5.4) this condition is

$$(T_{11})_{x=\pm h} = (c_{11}\partial u + d_{11}\partial P)_{x=\pm h} = 0. \tag{9.5b}$$

Finally, we suppose that the voltages applied to the dielectric at the interfaces are again

$$(\varphi)_{x=\pm h} = \pm V. \tag{9.5c}$$

We have now to solve (8.1) subject to the boundary conditions (9.5a,b,c). Let

$$\begin{aligned} u &= B_1 \cosh(x/\ell), \\ P &= A_2 + B_2 \cosh(x/\ell), \\ \varphi &= A_3 x + B_3 \sinh(x/\ell). \end{aligned} \tag{9.6}$$

Upon substituting (9.6) in (8.1), we find

(9.7) $\quad A_2 = \epsilon_0 \eta A_3 , \quad B_3 = \ell B_2/\epsilon_0 = - \ell c_{11} B_1 / \epsilon_0 d_{11} ,$

(9.8) $\quad \ell = \left(\dfrac{b_{11} c_{11} - d_{11}^2}{c_{11}(a_{11} + \epsilon_0^{-1})} \right)^{\frac{1}{2}} = \left(\dfrac{\epsilon_0 (b_{11} c_{11} - d_{11}^2)}{c_{11}(1 + \eta^{-1})} \right)^{\frac{1}{2}}$

i.e., ℓ is the same as in (8.5).

As for boundary conditions, (9.5b) is satisfied identically while (9.5a) and (9.5c) become

(9.9)
$$A_2 + B_2 \cosh(h/\ell) = - k\epsilon_0 \eta V/h ,$$
$$A_3 h + B_3 \sinh(h/\ell) = V ,$$

respectively. From (9.7) and (9.9),

(9.10)
$$A_3 = (B_3/\eta\ell) \cosh(h/\ell) + kVh ,$$
$$B_3 = (1-k)\eta V / \left[\eta \sinh(h/\ell) + (h/\ell) \cosh(h/\ell) \right].$$

The remaining constants are obtained easily from (9.10) and (9.7).

The capacitance, ignoring any voltage drop that may occur in the electrodes, is

(9.11) $\quad C = \dfrac{(\epsilon_0 \partial\varphi - P)_{x=\pm h}}{2V} = \dfrac{\epsilon}{2h} \dfrac{1 + (k\eta\ell/h)\tanh(h/\ell)}{1 + (\eta\ell/h)\tanh(h/\ell)} .$

In Fig. 9.2 the heavy, full line shows the relation between normalized inverse capacitance and normalized thickness, according to (9.11), for the case $k = 0.1$, $\eta = 10$. Calculations from Mead's data, for small k, indicate that the material property ℓ for all his dielectrics is of the order of magnitude

Variation of Capacitance with Thickness

of a few angstroms; and this is supported by Askar, Lee and Cakmak's calculations for alkali halides [11]. Hence, Mead's data, which do not extend below a thickness of 30 Å, would be well to the right of the knee of the curve, in Fig.9.2, and so would give the appearance of a linear relation which, if extended to zero thickness, would have a positive, non-zero intercept of inverse capacitance. This intercept, according to (9.11), is

$$C_0^{-1} = 2\ell(1-k)\eta/\epsilon. \qquad (9.12)$$

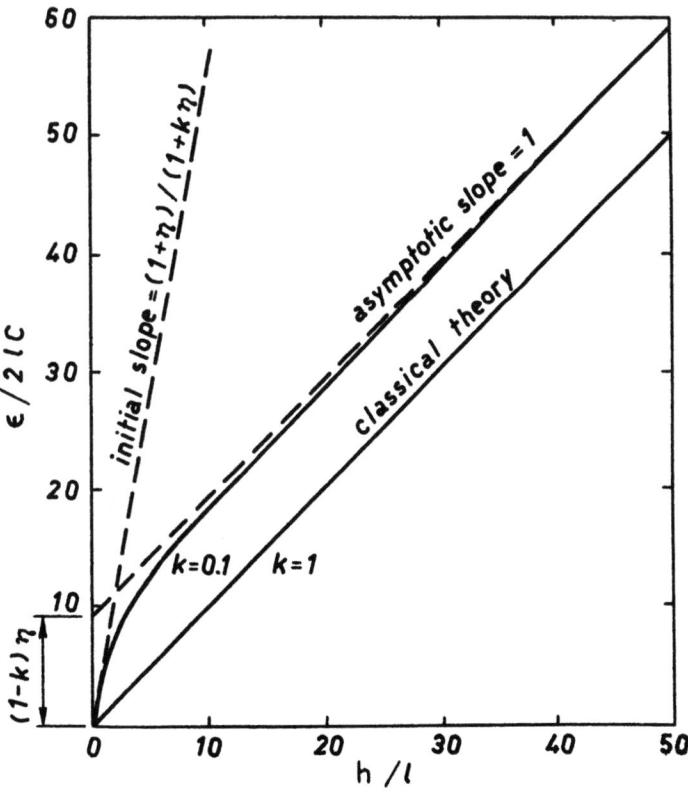

Fig.9.2

If $k = 1$, the intercept reduces to zero and, in fact, the whole solution reduces to the classical one. However, it seems unlikely that the presence of the metal would not influence the polarization of the dielectric at the metal-dielectric interface.

The variations of polarization and potential across the thickness of the dielectric are illustrated by the curves in Fig.9.3. The absolute value of the polarization is almost uniform across the major portion of the thickness and slightly less than the uniform polarization of the classical theory; but then drops sharply, near the surfaces, to boundary values of k times the classical polarization, as specified. The potential has an almost uniform gradient, less than the uniform gradient of the classical theory, over most of the thickness, but then increases sharply on approaching the boundaries.

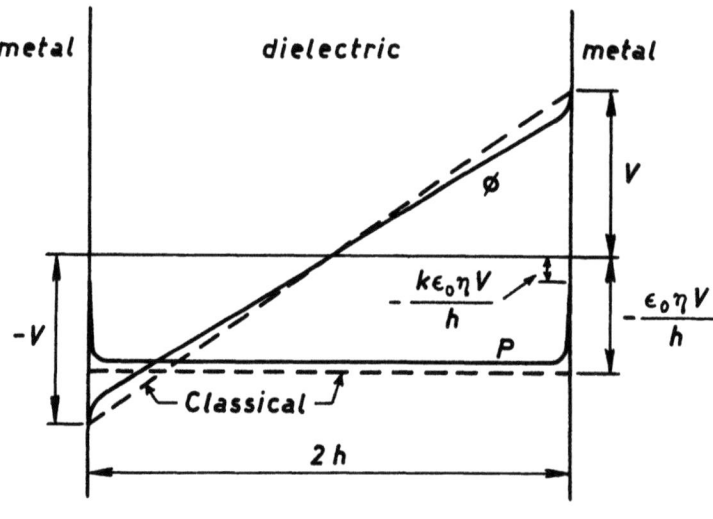

Fig.9.3

10. Lattice of Shell-Model Atoms.

In this section the equations of equilibrium are derived for a monatomic, one-dimensional lattice of the Cochran type [13] based on the Dick-Overhauser [12] shell-model of the atom: a core, comprising the nucleus and inner electrons, surrounded by a shell of outer electrons. The polarization is proportional to the relative displacement of the core and shell of the atom. In addition to this intra-atomic interaction, account is taken of interatomic interactions between core and core, core and shell and shell and shell of nearest neighbor atoms. It is shown that the equations of the lattice have, as their continuum limit, the equations of the augmented theory, including the contribution of the polarization gradient to the stored energy, rather than the classical theory of elastic dielectrics. The additional effects associated with the new constants b_{11} and d_{11} stem primarily from the shell-shell interaction. This interaction is known to be important in the matching of lattice dispersion relations to neutron diffraction dispersion data at short wave lengths [14,15].

We consider a single line of atoms extracted from a three-dimensional, monatomic lattice of shell-model atoms (Fig.10.1). The atoms are situated at $x = na$ where $\underline{n}$ is a positive or negative integer and $\underline{a}$ is the lattice constant, i.e. the distance between nearest neighbor atoms.

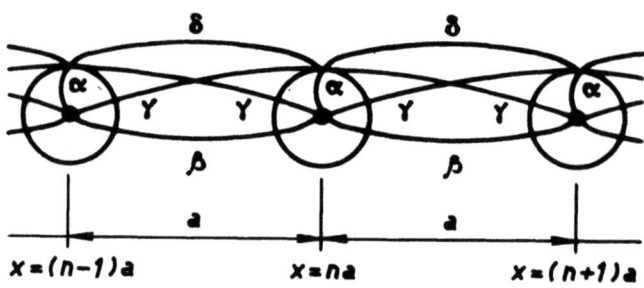

Fig.10.1

The displacements of the core and shell of the atom at $x = na$ are designated by u_n and s_n, respectively. The force constant, i.e. the force per unit relative displacement, of the intra-atomic core-shell interaction is designated by α and the force constants of the interatomic core-core, core-shell and shell-shell interactions are designated by β, γ and δ, respectively.

The equation of equilibrium of the nth interior <u>atom</u> (core and shell combined) is obtained by setting equal to zero the sum of the forces on its core and shell exerted by the cores and shells of its two nearest neighbor atoms:

$$\beta(u_{n+1} - u_n) + \gamma(s_{n+1} - u_n) + \gamma(u_{n+1} - s_n) + \delta(s_{n+1} - s_n)$$
(10.1)
$$- \beta(u_n - u_{n-1}) - \gamma(u_n - s_{n-1}) - \gamma(s_n - u_{n-1}) - \delta(s_n - s_{n-1}) = 0.$$

The <u>shell</u>, alone, of the nth atom is acted upon by its own core, the core and shell of each of the two neighboring atoms and, also, the Maxwell, electric self-field (which occupies all space). Hence,

$$\alpha(u_n - s_n) + \gamma(u_{n+1} - s_n) + \delta(s_{n+1} - s_n)$$
$$-\gamma(s_n - u_{n-1}) - \delta(s_n - s_{n-1}) + q E_n = 0, \qquad (10.2)$$

where q is the charge of the atom and E_n is the value, at $x = na$, of the Maxwell field. Now, $E_n = -[\partial \varphi]_{x=na}$ and this may be expressed in terms of φ_n, the value of the potential φ at $x = na$, by expanding the derivative, ∂, in an infinite series of forward differences [26]:

$$-E_n = \partial_+ \varphi_n = \sum_{m=1}^{\infty} (-1)^{m-1} m^{-1} a^{m-1} \Delta_+^m \varphi_n, \qquad (10.3)$$

where

$$\Delta_+ \varphi_n = (\varphi_{n+1} - \varphi_n)/a,$$
$$\Delta_+^2 \varphi_n = (\varphi_{n+2} - 2\varphi_{n+1} + \varphi_n)/a^2, \qquad (10.4)$$
$$\Delta_+^3 \varphi_n = (\varphi_{n+3} - 3\varphi_{n+2} + 3\varphi_{n+1} - \varphi_n)/a^3,$$
$$\vdots \qquad \vdots$$

The polarization of the nth atom, per unit area of the three-dimensional lattice, is defined by

$$P_n = (s_n - u_n) q/a^3. \qquad (10.5)$$

Further, let us adopt a symbol for the second central difference (divided by a^2):

(10.6) $\quad \Delta^2 f_n = (f_{n+1} - 2f_n + f_{n-1})/a^2.$

Then the equations of equilibrium (10.1) and (10.2) may be rearranged to the following forms:

(10.7) $\quad (\beta + 2\gamma + \delta)a^{-1}\Delta^2 u_n + (\gamma + \delta)a^2 q^{-1}\Delta^2 P_n = 0,$

(10.8) $\quad (\gamma + \delta)a^2 q^{-1}\Delta^2 u_n + \delta a^5 q^{-2}\Delta^2 P_n - (\alpha + 2\gamma)a^3 q^{-2} P_n - \partial_+ \varphi_n = 0.$

It will be observed that the difference equations (10.7) and (10.8) have exactly the same form as the first two of (8.1), which are the one-dimensional differential equations of equilibrium of the augmented theory of elastic dielectrics. Accordingly, if we set

(10.9) $\quad \begin{aligned} & a_{11} = (\alpha + 2\gamma)a^3 q^{-2} = \epsilon_0^{-1}\eta^{-1}, \quad b_{11} = \delta a^5 q^{-2}, \\ & c_{11} = (\beta + 2\gamma + \delta)a^{-1}, \quad d_{11} = (\gamma + \delta)a^2 q^{-1}, \end{aligned}$

and recall that [26]

(10.10) $\quad \Delta^2 = \partial^2 + \dfrac{a^2 \partial^4}{12} + \dfrac{a^4 \partial^6}{360} + \cdots,$

the first two of (8.1) are the lowest order continuum approximations to the equations of equilibrium of the atom and its shell, respectively.

As may be seen from (10.9), the polarization gradient terms (those with coefficients b_{11} and d_{11}) in (8.1) stem from the shell-shell and interatomic core-shell interactions, identified by the force constants δ and γ, respectively. In fact, the form of (8.1) is preserved if the interatomic core-shell interaction is omitted but the shell-shell interaction is retained. If both of these interactions are omitted, the continuum approximation reduces to (9.1), which is Toupin's form of the classical equations. The second of (9.1), Toupin's "equation of intramolecular force balance", does not appear in the traditional equations of elastic dielectrics; but we see that it is a fundamental equilibrium condition: the equilibrium of the shell under the action of the core of the same atom and the surrounding Maxwell field. The corresponding equation in the extended theory (the second of (8.1)) includes, in addition, the action of the adjacent atoms on the shell (in the continuum approximation).

We have yet to derive the lattice counterpart of the third of (8.1) and establish the boundary conditions. It is illuminating to reach these results from considerations of energy.

By analogy with (3.3) and (4.8), let us take, for the electric enthalpy of the one-dimensional lattice (per unit volume a^3)

$$\mathcal{H} = \sum_n \left[b_0 \Delta_+ P_n + \frac{1}{2} \epsilon_0^{-1} \eta^{-1} P_n^2 + \frac{1}{2} b_{11} (\Delta_+ P_n)^2 + c_{11} (\Delta_+ u_n)^2 \right.$$
$$\left. + d_{11} (\Delta_+ P_n)(\Delta_+ u_n) - \frac{1}{2} \epsilon_0 (\partial_+ \varphi_n)^2 + P_n (\partial_+ \varphi_n) \right]. \quad (10.11)$$

We have to find the derivatives of $\mathcal{H}$ with respect to u_n, P_n and φ. The first two are conventional and straightforward, for example,

$$\frac{\partial(\Delta_+ u_m)^2}{\partial u_n} = \frac{1}{a^2} \frac{\partial(u_{m+1}-u_m)^2}{\partial u_n} = \frac{2}{a^2}(u_{m+1}-u_m)\frac{\partial(u_{m+1}-u_m)}{\partial u_n},$$

$$= \frac{2}{a^2}(u_{m+1}-u_m)(\delta_n^{m+1}-\delta_n^m),$$

(10.12) $$= \frac{2}{a^2}(u_n - u_{n-1} - u_{n+1} + u_n) = 2\Delta^2 u_n,$$

where δ_q^p is the Kronecker delta.

To find the derivative of $\partial_+ \varphi$ with respect to φ, note first that, from (10.3),

$$\partial_+ \varphi_n = a^{-1}\left[\varphi_{n+1} - \varphi_n - \frac{1}{2}(\varphi_{n+2} - 2\varphi_{n+1} + \varphi_n) + \frac{1}{3}(\varphi_{n+3} - 3\varphi_{n+2} + 3\varphi_{n+1} - \varphi_n) - \ldots\right].$$

Then

$$P_m \frac{\partial(\partial_+ \varphi_m)}{\partial \varphi_n} = a^{-1} P_m \left[\delta_n^{m+1} - \delta_n^m - \frac{1}{2}(\delta_n^{m+2} - 2\delta_n^{m+1} + \delta_n^m) + \frac{1}{3}(\delta_n^{m+3} - 3\delta_n^{m+2} + 3\delta_n^{m+1} - \delta_n^m) - \ldots\right],$$

$$= a^{-1}\left[P_{n-1} - P_n - \frac{1}{2}(P_{n-2} - 2P_{n-1} + P_n) + \frac{1}{3}(P_{n-3} - 3P_{n-2} + 3P_{n-1} - P_n) - \ldots\right],$$

(10.13) $$= -\partial_- P_n,$$

Finite Difference Equations of the Lattice

where

$$\partial_- P_n = \sum_{m=1}^{\infty} m^{-1} a^{m-1} \Delta_-^m P_n \qquad (10.14)$$

and

$$\Delta_- P_n = (P_n - P_{n-1})/a,$$

$$\Delta_-^2 P_n = (P_n - 2P_{n-1} + P_{n-2})/a^2, \qquad (10.15)$$

$$\Delta_-^3 P_n = (P_n - 3P_{n-1} + 3P_{n-2} - P_{n-3})/a^3,$$
$$\vdots \qquad \vdots$$

i.e., $\partial_- P_n$ is the expansion of ∂P in an infinite series of backward differences [26]. Finally,

$$\frac{\partial (\partial_+ \varphi_m)^2}{\partial \varphi_n} = 2 \partial_+ \varphi_m \frac{\partial (\partial_+ \varphi_m)}{\partial \varphi_n} = -2 \partial_- \partial_+ \varphi_n. \qquad (10.16)$$

We now find the equations of equilibrium:

$$-\frac{\partial \mathcal{H}}{\partial u_n} = c_{11} \Delta^2 u_n + d_{11} \Delta^2 P_n = 0,$$

$$-\frac{\partial \mathcal{H}}{\partial P_n} = d_{11} \Delta^2 u_n + b_{11} \Delta^2 P_n - \epsilon_0^{-1} \eta^{-1} P_n - \partial_+ \varphi_n = 0, \qquad (10.17)$$

$$-\frac{\partial \mathcal{H}}{\partial \varphi_n} = -\epsilon_0 \partial_- \partial_+ \varphi_n + \partial_- P_n = 0.$$

The first two of (10.17) are the same as (10.7) and (10.8), with (10.9). This justifies the assumption (10.11) for the electric enthalpy. Also, we have found the third of (10.17), which has the third of (8.1) as its continuum form.

If the lattice is of finite thickness spanning an

odd number of atoms with the end ones at $n = \pm N$, the conditions for free boundaries are

$$-\frac{\partial \mathcal{H}}{\partial(\Delta_+ u_{\pm N})} = c_{11}\Delta_+ u_{\pm N} + d_{11}\Delta_+ P_{\pm N} = 0,$$

(10.18) $$-\frac{\partial \mathcal{H}}{\partial(\Delta_+ P_{\pm N})} = d_{11}\Delta_+ u_{\pm N} + b_{11}\Delta_+ P_{\pm N} + b_0 = 0,$$

$$-\frac{\partial \mathcal{H}}{\partial(\partial_+ \varphi_{\pm N})} = \epsilon_0 \partial_+ \varphi_{\pm N} - P_{\pm N} = 0,$$

where

(10.19) $$\Delta_+ f_{\pm N} = (f_{\pm(N+1)} - f_N)/a, \quad \partial_+ \varphi_{\pm N} = (\partial_+ \varphi_n)_{n=\pm N}.$$

In general, admissible boundary conditions are the specification of one member of each of the three products $u_{\pm N}(c_{11}\Delta_+ u_{\pm N} + d_{11}\Delta_+ P_{\pm N})$, $P_{\pm N}(d_{11}\Delta_+ u_{\pm N} + b_{11}\Delta_+ P_{\pm N} + b_0)$, $\varphi_{\pm N}(\epsilon_0 \partial_+ \varphi_{\pm N} - P_{\pm N})$; i.e. eight combinations in all, for the one-dimensional case.

11. Solutions of the Lattice Equations.

As an example of solutions of the lattice equations, consider the one analogous to that in Section 9 for the capacitance of thin dielectric films. In place of the equations of equilibrium (8.1) we now have (10.17) and in place of the boundary conditions (9.5) we now have

$$c_{11}\Delta_+ u_{\pm N} + d_{11}\Delta_+ P_{\pm N} = 0,$$

$$P_{\pm N} = -k\epsilon_0 \eta V/h, \qquad (11.1)$$

$$\varphi_{\pm N} = \pm V,$$

where $h = Na$. We take

$$u_n = B_1' \cosh(na/\lambda),$$

$$P_n = A_2' + B_2' \cosh(na/\lambda), \qquad (11.2)$$

$$\varphi_n = \left[A_3' x + B_3' \sinh(x/\lambda)\right]_{x=na},$$

and note that $\partial_\pm [f_n(na)] = [\partial f(x)]_{x=na}$ and

$$\Delta^2 \cosh \frac{na}{\lambda} = a^{-2}\left[\cosh \frac{(n+1)a}{\lambda} + \cosh \frac{(n-1)a}{\lambda} - 2\cosh \frac{na}{\lambda}\right],$$

$$= 4a^{-2} \sinh^2(a/2\lambda) \cosh(na/\lambda). \qquad (11.3)$$

Then, substituting (11.2) in (10.17), we find, analogous to (9.7),

(11.4) $A_2' = \epsilon_0 \eta A_3'$, $\quad B_3' = \lambda B_2'/\epsilon_0 = \lambda c_{11} B_1'/\epsilon_0 d_{11}$

and, in place of (9.8),

(11.5) $\sinh(a/2\lambda) = a/2\ell$,

where ℓ is the same as in (9.8). Thus (11.4) is the same as (9.7) with ℓ replaced by λ; but ℓ and λ are not the same: being related in accordance with (11.5).

Application of the boundary conditions (11.1) leads, by the same procedure as in Section 8, to

(11.6) $A_3' = (B_3'/\eta\lambda)\cosh(h/\lambda) + kV/h$,

$B_3' = (1-k)\eta V/\left[\eta \sinh(h/\lambda) + (h/\lambda)\cosh(h/\lambda)\right]$,

analogous to (9.10). Finally, the capacitance is

(11.7) $C = \dfrac{\epsilon_0 \partial_+ \varphi_{\pm N} - P_{\pm N}}{2V} = \dfrac{\epsilon}{2h} \dfrac{1 + (k\eta\lambda/h)\tanh(h/\lambda)}{1 + (\eta\lambda/h)\tanh(h/\lambda)}.$

Thus, the entire solution is identical with the continuum one except that the displacement and polarization have significance only at the atom sites and ℓ is replaced by λ. In particular, the curve in Fig. 9.3 is applicable to the lattice if ℓ is replaced by λ in both abscissa and ordinate.

A somewhat similar result is obtained for the strain and polarization at a free surface, analogous to the problem solved in Section 8. The main difference is that, owing to the geometric asymmetry of the half-space, there is a shift of **a/2** between the continuum and lattice solutions.

References.

[1] W. Voigt: Lehrbuch der Kristallphysik. B.G.Teubner, Leipzig, 2nd Edition, 1928; Edwards Brothers, Ann Arbor, 1946.

[2] W.P. Mason: Piezoelectric Crystals and their Application to Ultrasonics. D.Van Nostrand, New York, 1950.

[3] H.F. Tiersten: Linear Piezoelectric Plate Vibrations. Plenum Press, New York, 1969.

[4] R.D. Mindlin: Problems of Continuum Mechanics. J.R.M.Radok, Editor, Soc.Industrial and Appl.Math., Philadelphia, 1961, p.282.

[5] R.A. Toupin: J.Rational Mech.and Anal.Vol.5, 1956, pp.849--915.

[6] R.D. Mindlin: Int.J.Solids Structures, Vol.4, 1968, pp.637--642.

[7] M.P. Tosi: Solid State Physics Vol.16, Academic Press, New York, 1964, p.92.

[8] G.C. Benson and K.S. Yun: The Solid-Gas Interface. E.A.Flood, Editor, M.Dekker, Inc., New York, 1967, p.203.

[9] R.D. Mindlin: Int.J.Solids Structures, Vol.5, 1969, pp.1197--1208.

[10] C.A. Mead: Phys.Rev.Letters, Vol.6, 1961, p.545.

[11] A. Askar, P.C.Y. Lee and A.S. Cakmak: Phys.Rev.B, May 1970.

[12] B.G. Dick and A.W. Overhauser: Phys.Rev.Vol.112, 1958, p.90.

[13] W. Cochran: Proc.Roy.Soc., Vol.A 253, 1959, p.260.

References

[14] W. Cochran: Phonons in Perfect Lattices... R.W.H.Stevenson, Editor, Plenum Press, New York, 1966, p.53.

[15] W. Ludwig: Recent Developments in Lattice Theory. Springer Verlag, Berlin, 1967.

[16] Standards on Piezoelectric Crystals, 1949, Proc.Inst.Radio Engineers, Vol.37, 1949, pp.1378-1395.

[17] J.C. Slater and H. Frank: Electromagnetism. Mc Graw-Hill, N.Y., 1947.

[18] J. Schwartz: Int.J.Solids Structures, Vol.5, 1969, pp.1209--1220.

[19] J.S. Lomont: Applications of Finite Groups. Academic Press, N.Y.,1959.

[20] C.A. Mead: Phys.Rev.,Vol.128, 1962, p.2088.

[21] C.A. Mead and M.Mc Coll: Trans.Metall.Soc.A.I.M.E. Vol.233, 1965, p.502.

[22] C.A. Mead: Proc.Int.Symp.on Basic Problems in Thin Film Physics, Vandenhoeck and Ruprecht, Göttingen, 1966, p.674.

[23] H.Y. Ku and F.G. Ullmann: J.Appl.Phys., Vol.35, 1964, p.265.

[24] L.H. Germer, A.U. Mac Rae and C.D. Hartman: J.Appl.Phys., Vol.32, 1961, p.2432.

[25] A. Askar, P.C.Y. Lee and A.S. Cakmak: Int.J.Solids Structures, forthcoming.

[26] M.G. Salvadori and M.L. Baron: Numerical Methods in Engineering, Prentice-Hall, New Jersey, 2nd Edition, 1961.

Contents.

	Page
Preface	3
1. Introduction	5
2. Classical Equations	7
3. Toupin's Variational Principle for the Classical Theory	9
4. Polarization Gradient	16
5. Centrosymmetric Materials: Cubic	21
6. Centrosymmetric Materials: Isotropic	24
7. Surface Energy of Deformation and Polarization	26
8. Energy at a Plane, Free Surface	30
9. Capacitance of Thin, Dielectric Films	34
10. Lattice of Shell-model Atoms	41
11. Solutions of the Lattice Equations	49
References	53

If you have any concerns about our products,
you can contact us on
ProductSafety@springernature.com

In case Publisher is established outside the EU,
the EU authorized representative is:
**Springer Nature Customer Service Center GmbH
Europaplatz 3, 69115 Heidelberg, Germany**

Printed by Libri Plureos GmbH
in Hamburg, Germany